yes, we can!

She Can STEM

CHEMISTRY & PHYSICS

QUARRY

yes, we can!
She Can STEM

CHEMISTRY & PHYSICS

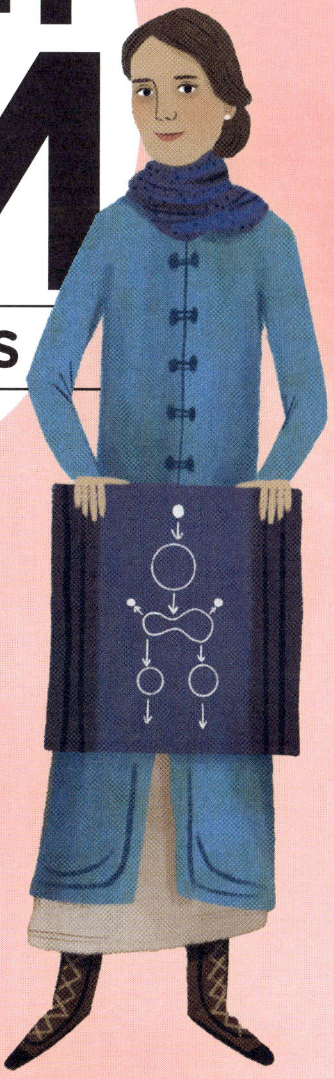

Inspiring Stories of Women Who Changed the World

LIZ LEE HEINECKE
The Kitchen Pantry Scientist

QUARRY

Quarto.com

This library edition published in 2026 by Quarry Books,
an imprint of The Quarto Group.

100 Cummings Center, Suite 265-D, Beverly, MA 01915, USA.
T (978) 282-9590 F (978) 283-2742

Distributed in the United States and Canada
by Lerner Publisher Services
241 First Avenue North
Minneapolis, MN 55401 U.S.A.
www.lernerbooks.com

First Library Edition

Library of Congress Cataloging-in-Publication Control Number: 2025032028

Printed in China

10 9 8 7 6 5 4 3 2 1

To each woman, famous or unseen,
who paved the way for women in science.

And to those treading the road today
to make a better future for all of us.

CONTENTS

INTRODUCTION 9

10 TAPPUTI-BELATIKALLIM Fragrance Distillation

12 AGNES POCKELS Surface Tension

14 MARIE CURIE Elemental Extraction

16 ALICE BALL Organic Separation

18 GERTY CORI The Cori Cycle

20 MARIA GOEPPERT-MAYER Nuclear Shell Model

22 ANNA JANE HARRISON Organic Compounds/ Ultraviolet Light

24 EDITH FLANIGEN Molecular Sieves

26 TU YOUYOU Medicinal Plant Compounds

28 ADA YONATH Ribosome Structure

30 MARGARET CAIRNS ETTER Crystallography

32 RAYCHELLE BURKS Colorimetric Sensors

Chemistry Physics

34

LAURA BASSI
Air/
Electricity

36

LISE MEITNER
Nuclear
Fission

38

KATHARINE BURR
BLODGETT
Thin Films

40

CECILIA
PAYNE-GAPOSCHKIN
Star Material

42

MARY GOLDA ROSS
Aerospace
Engineering

44

CHIEN-SHIUNG
WU
Symmetry

46

RUBY
PAYNE-SCOTT
Sunspots

48

CHRISTINE DARDEN
Aircraft Wing
Design

50

VALERIE L. THOMAS
Illusion
Transmitter

52

JOCELYN BELL
BURNELL
Pulsars

54

NADYA MASON
Conductivity/
Carbon

56

CHANDA
PRESCOD-WEINSTEIN
Cosmology

58

BURÇIN
MUTLU-PAKDIL
Galaxies

ABOUT THE AUTHOR 60
ABOUT THE ILLUSTRATOR 61
ACKNOWLEDGMENTS 62

CH3
H
O–O
H
H3C
O
CH3
H
O
O

8
O
(16)
84
Po
(209)
88
Ra
(226)

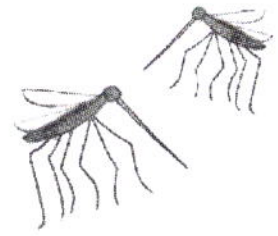

INTRODUCTION

When you think of a scientist, what image pops up in your mind? Do you picture an old man with crazy hair? That's how the world has been portraying scientists for a long time. Luckily, the face of science has changed dramatically and is continuing to evolve. Today, everyone should be able to see themselves as a scientist.

Women have been involved in STEM (Science, Technology, Engineering, and Math) for thousands of years series and today, there are more women than ever working in these disciplines. This book series lays out the biographies of 50 amazing women who are part of the scientific landscape, from historical scientists to modern ones. Many are role models who have inspired other girls and women to chase scientific dreams.

For example, in 1898, a physicist named Marie Curie became interested in invisible energy given off by some rocks. She figured out how to measure the energy and named it "radioactivity." Soon, she and her husband Pierre discovered two new radioactive elements and named them radium (Ra) and polonium (Po). Marie then invented a method to purify radium and won two Nobel prizes for her work, one in physics and one in chemistry. Few females did science at that time, but when girls and young women read about Marie's work, they realized that they could be scientists too.

Another scientist featured in this series helps us imagine how it might have felt to be the only woman in a large physics class in the 1960s. Day after day, Jocelyn Bell Burnell was harassed by fellow university students who stomped their feet, whistled, and banged on their desks as she walked through the lecture hall. In 1974, Jocelyn was the first person to discover spinning dead stars called radio pulsars, but she was left out when two men she worked with won a Nobel Prize for her discovery.

For centuries, women in STEM struggled to gain recognition for their work. In the 1700s, Laura Bassi was not allowed to lecture in public without the permission of men. In 1925, Cecilia Payne-Gaposchkin wrote a paper based on her research showing that the sun is made of hydrogen and helium. Two male astronomy professors told her she was wrong and then took credit for the discovery when they realized she was correct.

In 1957, the physicist Chien-Shiung Wu was ignored when a Nobel prize was awarded to two men whose theory she had confirmed by designing and performing a complicated physics experiment.

Despite the obstacles, women like these overcame significant hardships to make groundbreaking contributions to STEM fields. Their curiosity, passion, and perseverance made a difference. By reading about the modern scientists in this book, you'll see that, although things still aren't perfect, people who love science are no longer limited by their gender. Each biography also includes an activity, if you'd like to get a better feel for the science behind the stories.

Next time you picture a scientist, I hope you'll think of Raychelle Burks, Danielle Lee, Robin Wall Kimmerer, Lesley de Souza, Chanda Prescod-Weinstein, or one of the other brilliant scientists whose stories appear in these pages. Because once we move past the image of an old man with crazy hair, the sky is the limit.

Tapputi-Belatikallim

FRAGRANCE DISTILLATION

ROYAL PERFUMER

Over 3,200 years ago in ancient Mesopotamia, a pen made from a reed inscribed the name Tapputi-Belatikallim in a soft clay tablet. Today the cuneiform notches and figures remain, and scholars who can read the ancient script tell us that she held an important post as the manager of the royal household. Thanks to the scratches in baked clay, which held its form as the centuries rolled by, we understand that Tapputi, whose second name means "overseer," prepared fragrances for the king and his family. The royals used some of the scents she prepared as perfume and saved others to offer to the gods during religious rituals.

MESOPOTAMIA

Ancient Mesopotamia lay between two rivers, the Tigris and the Euphrates, in a land known as the Fertile Crescent where Iraq and parts of Iran, Syria, and Turkey are found today. The fertile land gave birth to the first known cities: agriculture arose on the land, people began to domesticate animals, and written language developed. Mesopotamia was a center for culture and language and gave rise to the wheel, the chariot, and the writings of a high priestess named Enheduanna, the world's first author known by name.

TRANSCENDENT SCENTS

When Tapputi made her mark as the world's first recorded chemist around 1200 BCE, tendrils of fragrance were intertwined in social order, religion, and medicine. Tapputi and her contemporaries believed that the invisible, beautiful scents they offered could transcend the physical world to reach their gods, who would be pleased by their sacrifice. Kings were self-proclaimed conduits to the gods, anointed with the most valuable perfumes, while fragrant ointments and salves were used by healers as well.

ROYAL RECIPE

Tapputi's recipe for a perfume included preparing a mixture of oil, flowers, and a lemon-grass-type herb called calamus and then steeping the mixture with other fragrant substances before filtering and distilling it again and again. The distillates and oils she created served as salves and perfumes for the king. Her complicated, multi-step extraction method was recorded around 1200 BCE. An ancient copy of her recipe includes the first description of a distillation apparatus ever recorded and a number of her methods are still used today in modern perfume production.

IN TODAY'S WORLD

Chemical engineers use several of the same techniques that Tapputi used, but on a much larger scale, to create today's fragrances.

TRY THIS AT HOME!

Crush a fragrant flower. Breaking the plant cells open releases scent-laden molecules called essential oils. These oils are volatile, which means they quickly evaporate into the air.

Agnes Pockels

SURFACE TENSION

KITCHEN SINK CHEMIST

Agnes Pockels was born in Venice, Italy, in 1862. Her father was an officer in the royal Austrian army, but when she was nine years old, he became very ill with malaria. Her family moved to Brunswick in Lower Saxony, which is part of Germany. Although she attended a high school for girls and was fascinated by science, women were not accepted into universities. Agnes wanted nothing more than to continue her studies, but she was forced to remain at home to take care of her two ailing parents while her brother went off to college.

KITCHEN SINK SCIENCE

Fortunately for the field of chemistry, Agnes's curiosity was unstoppable. She began to notice interesting phenomena while washing dishes in her kitchen sink. Fascinated by how oils and particles formed films on water, she observed that those films could be disturbed by soaps and other materials. Agnes transformed her kitchen into a research station and dove into the study of surface tension, which is the name for the way molecules stick together on the surface of a fluid. Her brother Fredrich, who was studying science at the university, recognized his sister's hunger for knowledge and supported her in every way he could. Besides observing her work, he gave her access to a physics journal so that she could learn what other scientists were doing in their more well-equipped laboratories.

SLIDE TROUGH

By the time she was twenty, Agnes had invented a "slide trough," which allowed her to play with the way a liquid's surface behaved, by sliding a wire or metal strip across the top of it. She used her apparatus to create very thin layers of fluids and then to test the effect of different contaminants, such as fine powders, on surface tension. Besides discovering methods for applying uniform layers of particles to a surface, she discovered that her trough could sweep a surface free of surface contaminants.

PUBLICATION

Agnes shared her results by writing a letter to another scientist who studied the surface of liquids. He was impressed, and eventually her work was published in the famous journal *Nature*. She was delighted to learn that other scientists were using her research results and trough design in their own laboratories. Pockels continue studying surface tension, and her work led to many discoveries and innovations in surface tension research and material science.

A KIND HEART

Agnes Pockels's brother died in 1913, and following World War I her beloved physics journal was no longer published. She lived out the rest of her life in relative obscurity, always helping others. Four years before her death, Agnes was awarded the Laura R. Leonard Prize of the German Colloid Society, "for her quantitative investigation of the properties of interfaces and surface films, and for the methods she used, which have since become fundamental in modern colloid science."

IN TODAY'S WORLD

A small amount of detergent is added to toothpaste to reduce the surface tension of the mixture. This allows the active ingredients, such as fluoride, to spread across teeth during brushing.

TRY THIS AT HOME!

Cover the bottom of a dinner plate with milk and add a few drops of food coloring to the milk without stirring. Add a drop of dish soap to the milk to break the surface tension of the liquid.

88
Ra
(226)
84
Po
(209)

Marie Curie

ELEMENTAL EXTRACTION

RADIOACTIVITY PIONEER

Marie Curie, born Manya Sklodowska, entered the world in 1867 surrounded by a loving family in a house filled with laughter, books, and her father's scientific equipment. Although her childhood in Poland was filled with magical moments, challenges loomed from the start. Their lives were overshadowed by an oppressive Russian regime, which forced her father out of his academic job and the family into an apartment where they opened a boarding school to pay the bills. One of the boys who arrived to live with the family carried typhus fever, which killed Marie's beloved sister Zosia. Two years later, tuberculosis took her mother's life.

POLAND

The serious, gray-eyed girl took refuge in her education. By the time she was seventeen, she was actively involved in a secret "floating university," where she began to learn about chemistry and biology and dove into literature and culture. She came to believe that education was the only way to fight oppression. When she went to work as a nanny for a wealthy family, she continued to tutor the poor neighbor children, teaching them to read. Eventually she earned enough money to join her sister in Paris in 1891 to pursue her dream of attending the famous Sorbonne University.

PARIS

Marie fell in love with math and physics and graduated at the top of her class at the Sorbonne, despite being so preoccupied with her studies that she often forgot to eat. While searching for a home for some laboratory equipment, she met Pierre Curie, the man who would become her partner in science and in life. Together they embarked on an extraordinary adventure.

STRANGE RAYS

While pursuing her doctorate degree, Marie decided to study the strange new rays that a scientist named Becquerel had discovered emanating from the element uranium. Searching for other elements and minerals naturally emitting these rays, she discovered that uranium mining waste called pitchblende gave a stronger signal than uranium did on its own. By chemically separating the mining waste, she discovered two new elements: polonium and radium.

It took her four years to purify a sample of radium smaller than a grain of rice, but it was worth it when her treasured element took its place on the periodic table and she coined the word "radioactive." She could have made a small fortune by patenting her extraction method, but Marie believed that everyone should benefit from scientific discovery, so she freely shared her technique with the scientific community and industry.

TWO NOBEL PRIZES

Marie Curie went on to become the first woman to win a Nobel Prize in Physics, and later she went on to win a second Nobel in chemistry. Her curiosity, hard work, and determination changed the world, and although radioactivity from her work shortened her life, she continues to inspire scientists today.

IN TODAY'S WORLD

Marie Curie helped to create the first radiation therapy to target and kill cancer cells. Although radium is no longer used for radiation therapy, the technique is still used today to treat certain cancers.

TRY THIS AT HOME!

Most radioactive materials are considered too dangerous to have in homes, but you can learn about them online. Look up some images of radium, one of the elements discovered by Marie Curie.

Alice Ball

ORGANIC SEPARATION

FAMILY OF ARTISTS

Born in 1892 in Seattle, Alice Ball was exposed to chemistry at a very young age. Her grandfather was a famous photographer and one of the first people in the United States to take daguerreotype images, which required the use of iodine, copper, and mercury. Sadly, her grandfather was ill, possibly as a result of chemical exposure, and Alice's family moved to Hawaii for a year in hopes that the sunny climate and sea air would improve his health. When he died, they returned to Seattle, where Alice received degrees in pharmaceutical chemistry and pharmacy from the University of Washington.

HAWAII

For graduate school, Alice chose to attend the University of Hawaii, where she became the first woman and the first African American to earn a master's degree in chemistry at the school. For her graduate thesis, Alice separated out the chemical parts of kava root, in order to discover its active components. Harry Hollman, an assistant surgeon at a nearby hospital, heard about her work on kava root and asked her to help solve the problem of treating Hansen's disease, which is also called leprosy. A leper colony had been established on the Hawaiian island of Molokai in 1865, and thousands of Hawaiians with the disease had been sent there in an attempt to stop the spread of the disease.

A NEW TREATMENT FOR LEPROSY

As a professor at the University of Hawaii, Ball devoted every moment that she wasn't teaching to synthesizing a better treatment for the disfiguring disease. At the time, the only way to reduce the symptoms of Hansen's disease was to inject or swallow chaulmoogra oil, which had been extracted from the seeds of a type of evergreen tree. Unfortunately, the oil didn't work well. Painful injections formed a necklace of blisters under the skin that weren't easily absorbed by the patient's bodies, and the oil tasted so horrible it was impossible to drink without vomiting. In only a year, twenty-three-year-old Professor Ball figured out how to separate the oil into parts until she found a compound that could be extracted and made into an easily absorbable medicine for victims of Hansen's disease.

STOLEN WORK

Tragically, Alice Ball died when she was only twenty-four years old, soon after making her discovery. She may have died from exposure to chlorine gas during a lab demonstration, but it is unclear. Her death certificate was altered to state that she died of tuberculosis. The university's president, who was also a chemist, took over Ball's studies, publishing her results under his own name, and producing large quantities of the medicine she'd created.

AN ALLY

When he realized what had happened, Harry Hollman made it public that credit for Alice Ball's discovery had been stolen by the president of the university. Hollman published a paper of his own, crediting Alice for her work and dubbing her leprosy treatment the "Ball Method." Until effective antibiotics were invented in the 1940s, Alice Ball's treatment was the only effective medicine for Hansen's disease.

IN TODAY'S WORLD

Recently the University of Hawaii erected a plaque in her honor, awarded her a medal, and named a scholarship after her. The state of Hawaii now celebrates "Alice Ball Day." Organic separations processes similar to the methods used by Dr. Ball are still used in the pharmaceutical industry today.

TRY THIS AT HOME!

Use a garlic press to squeeze organic compounds from a chopped orange or lemon peels. Collect the oil floating on top of the juice and use it to flavor food or add scent to home-made soap or candles.

Gerty Cori

THE CORI CYCLE

SETTING SAIL

Named after a transatlantic ship, Gerty Theresa Cori was born in Prague in 1896. Her father was a chemist who directed a sugar refinery, and she was tutored at home by her mother until she was ten and began attending the local school for girls. By the time she was sixteen, Gerty had decided that she wanted to be a medical doctor but discovered that she lacked many of the classes that were required. She worked extremely hard and over a year learned the science, language, and math she would need to study medicine. Her uncle, who was a professor of pediatrics, encouraged her to apply to medical school and she was admitted to a university in Prague.

A LAB PARTNER

While she was in medical school, Gerty met Carl Cori, who thought that she was smart and funny. They both loved the outdoors and working in a laboratory. After finishing medical school, they married and Gerty went to work in a laboratory at a children's hospital, where she did research on thyroid disease in children. In 1922, they moved to the United States, where they hoped Gerty wouldn't be discriminated against on the basis of her Jewish heritage. They became U.S. citizens and went to work together at what is now the Roswell Park Comprehensive Cancer Center in New York.

THE CORI CYCLE

A director at the Roswell Research Institute, who didn't like having women in the laboratory, tried his best to get rid of Gerty, but Carl stood up for her and she refused to leave. Despite the hostile environment, the Coris worked together day and night to study how the human body turns carbohydrates into energy. In 1929, they proposed a "cycle of carbohydrates" in which a sugar called glucose in muscles is cleaved into lactic acid, which the liver can turn back into glucose for muscles to utilize again. Gerty was listed as the first author on the papers related to this work, and today that cycle is named after her and Carl. Eventually, the Coris moved to Washington University in St. Louis, where they continued their research. Gerty returned to her love of pediatric medicine by studying metabolic disorders in children.

A NOBEL PRIZE

In 1945 Gerty Cori became the first American woman, and the third woman ever, to win a Nobel Prize in science when she and Carl were awarded the Nobel Prize for Physiology or Medicine for their work on the Cori cycle. Sadly, like Marie Curie and Irene Curie who also won Nobel prizes, Dr. Cori's early death was probably the result of radiation exposure. Her contributions to science are foundational to scientific research that continues today.

IN TODAY'S WORLD

Dr. Cori's work on the Cori cycle is important for modern research on a disease called diabetes. Diabetes is a treatable disease that affects many people, and scientists are working hard to find a cure.

TRY THIS AT HOME!

Line up three clear jars or glasses. Label the end jars "muscle" and the center jar "liver." Fold and trim two paper towels to form bridges between the jars, reaching to the bottom. Fill one end jar with water and add a few drops of food coloring. The water will move from jar to jar via capillary action, representing how the body moves lactic acid waste from muscles to the liver, where it is turned into a sugar called glucose and then returned to muscles.

Maria Goeppert-Mayer

NUCLEAR SHELL MODEL

AN EDUCATION

Atoms are the building blocks of matter. Maria Goeppert-Mayer is best known for discovering how the spinning of protons and neutrons inside of atoms affects the structure of the nucleus and the stability of different elements. Born in 1906, Goeppert-Mayer grew up in Germany admiring her father, a sixth- generation professor of pediatrics who understood the importance of educating women. Although the preparatory school for girls she attended closed, Maria and four other girls continued to study and were able to pass their exams for acceptance into the university.

JUMPING ELECTRONS

Germany needed teachers for girls' schools and Maria Goeppert had planned to study math in college, but she found herself irresistibly drawn to physics and pursued a PhD in that discipline. In her doctoral paper, she described a theory called two-photon absorption that correctly predicted that bundles of energy called photons can make electrons jump to higher energy levels.

COMING TO AMERICA

In 1930, Goeppert married another scientist named Joseph Mayer. They moved to Baltimore, Maryland, in the United States, where Joseph had been offered a job at Johns Hopkins University. Maria was not offered a job, because there was an outdated rule against hiring spouses, which had become an excuse not to hire women as professors. They gave her a job writing letters in German, and in return she got a tiny salary along with access to a lab where she could continue her research. She also got to teach a few classes but was not paid.

PHYSICS "FOR THE FUN OF IT"

In 1937, Joseph went to work at Columbia University and Maria was given an office but no salary. She continued her research for the sheer pleasure of doing physics.

MANHATTAN TO CHICAGO

Shortly after beginning her first paid teaching position at Sarah Lawrence College, Dr. Goeppert-Mayer was recruited to work on the Manhattan Project, which eventually produced the world's first atomic bombs and brought World War II to an end. She and Joseph moved to Chicago, where she was offered a part-time job as a senior physicist at the nearby Argonne National Laboratory.

A WALTZ

It was in Chicago that Goeppert-Meyer did her Noble Prize–winning research on the structure of the nuclei of atoms. She liked to describe nuclei as concentric circles of waltzing couples, some spinning clockwise and others counterclockwise while the couples twirl "round and round." In 1960 Maria Goeppert-Meyer finally became a full professor. In 1963 she shared a Nobel Prize with J. Hans D. Jensen and Eugene Paul Wigner for her discoveries concerning nuclear shell structure, becoming the second woman in history to win a Nobel Prize in Physics.

IN TODAY'S WORLD

Today, astrophysicists are using the nuclear shell model to understand the science of exploding stars called supernovas.

TRY THIS AT HOME!

Use an orange or a Styrofoam ball to represent the nucleus of an atom. Use materials such as wire, wooden skewers, and beads to make a Bohr model of an element on the periodic table. Each element's atomic number will tell you how many electrons orbit the nucleus, and you can look up their arrangement in a book or online. Protons and neutrons are found inside the nucleus.

Anna Jane Harrison

ORGANIC COMPOUNDS/ULTRAVIOLET LIGHT

ONE-ROOM SCHOOLHOUSE

Anna Jane Harrison grew up on a farm in Missouri. When she asked her father about caterpillars, he taught her about the tractors instead of the butterfly larvae. She tucked the memory away as her earliest brush with science. When he died in 1919, her mother continued running the farm to support seven-year-old Anna and her brother. Anna loved going to school in a one-room schoolhouse and was inspired by her wonderful teacher there.

TEACHING

She attended college and graduate school at the University of Missouri, studying chemistry and education, and she taught at the one-room schoolhouse that she had attended as a child. In 1940, she finished her PhD in physical chemistry. She went to work teaching chemistry at a women's college in New Orleans.

SMOKE DETECTOR

During World War II, Anna took a break from teaching and went to work doing research for the government. In a lab in Kansas City, Missouri, she studied toxic smoke for the National Defense Research Committee and did some work at Corning Glass Works in New York. Some of the work she did resulted in the invention of smoke detectors.

LIGHT WORK

Dr. Harrison joined the faculty at Mount Holyoke College in Massachusetts to work with Emma P. Carr, an expert on ultraviolet spectroscopy, a laboratory technique that uses light waves to analyze chemicals. In addition to using spectroscopy to study chemicals, Harrison used a technique called flash photolysis to break organic chemicals apart.

As the reaction happened, she used spectroscopy to see what chemical fragments were created and whether they interacted with each other. She was especially interested in how organic compounds, which contain carbon linked to other elements, interact with ultraviolet light.

PUBLIC EDUCATION

Anna's students loved her. Besides having a good sense of humor, she was able to communicate complicated concepts in terms that everyone could understand. Dr. Harrison felt that it was essential for the public and policy makers to have a basic understanding of science, so that they could make decisions that would benefit everyone.

TRAVELING FOR SCIENCE

In 1978 Anna Harrison became the first female president of the American Chemical Society, and in 1983 she was named president of the American Association for the Advancement of Science. Her work gave her the opportunity to travel around the world on a mission to improve the way scientists talk to each other and communicate with the public. Dr. Harrison died in Massachusetts in 1998 when she was eighty-five years old.

IN TODAY'S WORLD

Today, organic photochemistry makes it easier for chemists to produce complex organic molecules for industry and research.

TRY THIS AT HOME!

On a baking sheet, arrange a stencil or some leaves on a piece of inexpensive, colorful construction paper. Cover with plastic wrap, weigh the wrap down with rocks, and place it in direct sunlight for several hours. Remove the plastic wrap and leaves or stencil to see how the chemicals in the paper fade in the Sun's ultraviolet light.

Edith Flanigen

MOLECULAR SIEVES

A MARVELOUS MENTOR

Edith Flanigen, who was born in 1929 in Buffalo, New York, credits her love of chemistry to a high school teacher. Edith recalls, "She really made (science) exciting. We did hands-on laboratory work, and I think I fell in love with chemistry at that time." Inspired by their teacher, Edith and her two sisters graduated from high school and went on to study chemistry at a nearby college. Following graduation, Edith and her sister, Joan, went on to get master's degrees in chemistry from Syracuse University.

CHEMICAL SPONGES AND EMERALDS

The early 1950s found Flanigen and her sisters working at Union Carbide Corporation, despite the fact that there were very few women working in chemistry at the time. Flanigen focused her research on identifying and purifying silicone polymers, which are compounds made up of repeating patterns of similar chemical units. Eventually she moved into a lab that studied molecular sieves, which are like chemical sponges that can mop up small particles while keeping large ones out. Besides working on molecular sieves, Flanigan co-invented synthetic emeralds that were manufactured by Union Carbide and used in pre-laser technology and in jewelry for a few years. She loved the creative, collaborative environment in industry and helped to foster those ideals in the labs where she worked.

ZEOLITE

In 1973, Flanigen became the first woman named Corporate Research Fellow at Union Carbide, and she eventually reached the highest possible technical positions that could be achieved at the corporations where she worked. During her career, she invented and co-invented more than 200 synthetic materials. Her name appears on more than 100 patents, which protect new inventions from being copied and sold by someone other than the inventor. Flannigan's research on molecular sieves was especially important. Filtering molecules (groups of atoms) by size allows scientists and industries to separate chemicals compounds. Edith's work with sieve materials called zeolites made it possible to refine oil into its usable parts more safely and efficiently.

PRESIDENTIAL RECOGNITION

In 1983, Flanigen received a PhD from D'Youville College. Besides being inducted into the National Inventors Hall of Fame in 2004, Dr. Edith Flanigan has received countless awards for her work. In 1992 she became the first woman ever awarded the Perkin Medal, and in 2014 President Barack Obama presented her with the National Medal of Technology and Innovation.

IN TODAY'S WORLD

Dr. Flanigen is the foremost authority on zeolite chemistry, and her work is the basis of the molecular sieve industry, which produces chemicals we use every day, including laundry detergent and cat litter.

TRY THIS AT HOME!

Outside, pour a mixture of sand and gravel into a kitchen sieve. Use a hose to rinse the mixture. Notice how sand and small stone pass through the sieve, while larger ones are trapped. This is a model of how molecular sieves can be used to trap and separate certain chemical compounds.

CH3
H
H3C
O–O
O
H
H
O
CH3
O

Tu Youyou

MEDICINAL PLANT COMPOUNDS

AN INTERRUPTED EDUCATION

Tu Youyou was born in a city on the east coast of China into a family that put a high value on education. The only girl in her family, she attended the best schools in the region until she became sick with tuberculosis when she was sixteen. By the time she had recovered and returned to school two years later, Tu had decided to study medicine so that she could stay healthy and help others. After graduating from high school, she was accepted into the school of pharmacy at the Medical School of Peking University.

PLANTS AS MEDICINE

In her pharmacy courses, Tu learned to study traditional Chinese remedies through the lens of Western medicine. Her professors taught her to identify and classify plants based on their botanical descriptions along with methods for extracting chemical compounds from them. Tu also learned how to characterize the structures of the chemicals she'd isolated. This education helped her to understand how some traditional plants are able to cure disease. Her first research project involved studying the effect of an herb called *Lobelia chinensus* on a disease caused by a parasitic flatworm.

A BROKEN FAMILY

Dr. Youyou left her two young daughters in the care of relatives so she could travel to a laboratory to search for a cure for malaria. Malaria is a deadly disease caused by a parasite spread by mosquitos, and the parasite had become resistant to a number of available medicines. Youyou searched through ancient Chinese texts and folk recipes, searching for new compounds to test against the malaria parasite. She collected more than 2,000 prescriptions to test and went to work experimenting with them.

A CURE

After a number of failures, an herb in the *Artemisia* family showed promising results, but she couldn't repeat them consistently. Going back through old books, she found a recipe that contained a cure for malaria fever that called for "A handful of Qinghao immersed in two liters of water, wring out the juice and drink it all." Dr. Youyou realized that most of the recipes had called for boiling the plant, which may have destroyed the chemical capable of killing the malaria parasite. She followed the heat-free recipe and it killed malaria parasites, so she isolated the fraction containing the active chemical. Due to limited resources, she and two of her colleagues tested the toxicity of the compound by drinking it themselves and found it to be safe enough for clinical trials.

A PRIZE

In 2015 Tu Youyou won the Nobel Prize in Physiology or Medicine for her discovery of artemisinin, a new therapy for malaria.

IN TODAY'S WORLD

Dr. Tu Youyou's discovery has saved millions of lives and is still used today to treat malaria all around the world.

TRY THIS AT HOME!

Peel off the bottom of an aloe vera leaf. Use a spoon to remove the gel inside. Refrigerate the aloe vera gel in a jar and use it to sooth irritated skin and sunburn.

Ada Yonath

RIBOSOME STRUCTURE

FALLING FOR SCIENCE

Ada Yonath set the stage for her scientific career early on. During an ill-conceived childhood experiment, she broke her arm falling from a stack of furniture she was using to measure the height of a balcony. Ada also recalls being inspired at a young age by a biography of Marie Curie.

A STRUGGLE

Ada Yonath's parents were from Poland, but they had left their home country to move to Jerusalem, where they struggled with poverty. Her father suffered from life-threatening health problems, but he and her mother had big dreams for Ada and did everything possible to ensure that she got a good education. After he died when she was only eleven, Ada and her mother worked for a year trying to make ends meet before moving to Tel Aviv around 1951 to be closer to relatives who could help them.

INTRODUCTION TO CRYSTALLOGRAPHY

In college Ada studied chemistry, biochemistry, and biophysics before going on to work toward her PhD at the Weismann Institute of Science. For her doctorate research, she studied the structure of a protein called collagen using X-ray crystallography, which bombards crystals with X-rays to determine their structure. After being awarded her PhD, she did postdoctoral research in the United States before returning home to establish the first biological crystallography laboratory in Israel.

MOUNT EVEREST

In the late 1970s, Dr. Yonath decided to tackle a monumental project. She wanted to discover the three-dimensional structure of ribosomes, which are the cellular machines that translate genetic material called RNA into proteins. Ribosomes are complex structures made up of two parts called subunits, and although many scientists told her that what she'd proposed couldn't be done, Ada was determined. Later, she compared her research journey to "climbing Mt. Everest only to discover that a higher Everest stood in front of us."

POLAR BEARS AND RIBOSOMES

To study the structure of ribosomes, first Dr. Yonath had to discover how to crystallize them. After reading about how the ribosomes in polar bears become neatly stacked during hibernation, Dr. Yonath got the idea to isolate organized ribosomes from bacteria that live in extremely cold, hot, or radioactive environments. She figured out how to flash freeze these crystallized ribosome structures and bombard them with X-rays. In 2001 Ada Yonath was awarded the Nobel Prize for discovering the structure and function of the ribosome.

IN TODAY'S WORLD

Today, Dr. Yonath's work has proved essential for the production of new antibiotics.

TRY THIS AT HOME!

Ribosomes are blob-like structures that read code from a molecule called RNA to make amino acids, which are the building blocks of proteins. Thread ring-shaped candy or cereal on a piece of yarn to see how ribosomes assemble chains of amino acids. Once they exit ribosomes, amino acid chains are folded into three-dimensional proteins.

Margaret Cairns Etter

CRYSTALLOGRAPHY

SCIENCE WITH A SMILE

Born in 1943 and nicknamed "Peggy," Margaret Carins Etter loved people and science. She especially liked people who do science. Her students, friends, and co-workers recall her laugh echoing down hallways and the smiley-face stickers she jokingly gave colleagues when they got good results in the lab. Etter was famous for always doing what she considered to be the right thing, even if it meant risking her career, and she made enormous contributions to science in a relatively short amount of time.

10,000 LAKES

Peggy Etter was born in Delaware, and her father was a chemist at DuPont Experimental Station. She grew up surrounded by scientists who worked with her dad, and she went to the University of Pennsylvania to study chemistry. After earning a master's degree at the University of Delaware, Peggy moved to Minnesota, the Land of 10,000 Lakes, where she earned a PhD and settled down to teach for a year at Augustana College before going to work at 3M.

THE BIG PICTURE

Eventually, Dr. Etter returned to the University of Minnesota where she became a full professor and dove into more research. She made important discoveries by studying hydrogen bonding patterns in chemical structures, rather than focusing on individual bonds. Etter was also fascinated by crystal growth and was one of the first scientists to observe and investigate the strange phenomenon of "jumping crystals," which are now studied in a field of chemistry called mechanochemistry.

AN ADVOCATE

As she explored the mysteries of the unknown on her laboratory bench, Etter fought against oppressive traditions in the worlds of industry and academia. According to her colleagues, Etter believed that actions spoke louder than words and wasn't afraid to push the boundaries. She spoke at schools to encourage girls to get into science, helped to establish the 3M Visiting Women's Scientist program, and was a director and instructor in their STEP program, which encourages minority students to become scientists.

A LASTING IMPRESSION

Although Dr. Etter died when she was only forty-nine, she left the chemistry world a better place. Her former students, colleagues, and close friends wrote a number of tributes to her, recalling her "intelligence, elegant presentation, enthusiasm, and radiant personality" and the fact that "she demonstrated what it meant to show respect and compassion for everyone, irrespective of status or behavior."

IN TODAY' S WORLD

Each year, the University of Minnesota holds an annual Etter Memorial Lecture and the American Crystallographic Association gives two awards in honor of Dr. Etter.

TRY THIS AT HOME!

With parental supervision, dissolve three cups (720 g) of Epsom salts in two cups (480 g) of water in the microwave. Cool slightly and pour into three jars. Make three small shapes from pipe cleaners (chenille sticks) and hang each one from a pencil positioned across the jars' mouths so they are suspended in the Epsom solution. Watch the crystals grow and remove the pipe cleaners when they're covered with needle-like salt crystals.

Raychelle Burks

COLORIMETRIC SENSORS

CALIFORNIA KID

Growing up in California, Raychelle Burks loved going to the library and solving mysteries alongside the sleuths of her favorite detective stories. Her family spent hours having long discussions about books and movies, which fueled her curiosity about popular culture. As she grew, their conversations around television shows such as *Star Trek* inspired her to think critically about what she observed in the world around her.

SUPER SLEUTH

Raychelle's grandmother had introduced her to Agatha Christie, and she was drawn to the observant characters who assembled information in an orderly way to piece together a possible scenario. When she was a teenager, a visit to a forensic science lab showed her how scientific equipment and chemical tests could be used to lead investigators to discoveries about crimes. She realized that science was a vehicle that could turn her passion for solving mysteries into a career.

REAL-LIFE TRICORDER

After pursuing her education, Dr. Burks spent a few years working in a crime lab before returning to academia. Today, she is a professor of chemistry who studies colorimetric sensors for chemical identity and concentration determination. She uses forensic science research to help reveal clues, doubling up as a science communication superhero when she's not busy with molecular sleuthing in her laboratory.

AWARDS

Dr. Burks was awarded the 2019 Mindlin Foundation Science Communication Prize and the 2020 American Chemical Society James T. Grady-James H. Stack Award for Interpreting Chemistry for the Public. In 2019 she was selected as an AAAS If/Then Ambassador by the American Association for the Advancement of Science and the Lyda Hill Philanthropies.

SCI-COMM SUPERHERO

As a science communicator and role model for aspiring scientists, Dr. Burks uses popular culture as a positive space for discussions about society, science, and stereotypes. In the classroom, online, and beyond, she works to ignite appreciation of chemistry through hands-on projects and pop culture references. Burks helped create and organize *SciPop Talks!*, a popular talk series blending science and pop culture. She appears regularly on television, podcasts, and at genre conventions, and she owns a cheweenie, which is a cross between a dachshund and a chihuahua.

IN TODAY'S WORLD

The chemical sensor Dr. Burks is working to develop in her lab is astonishingly similar to the fictional tricorder on *Star Trek*, which is also a hand-held device used to scan the environment. Her current research efforts are focused on the design of sensing systems that can identify chemical clues tied to crime and uses smartphones, laptops, and tablets as scientific analytical devices.

TRY THIS AT HOME!

Download a colorimeter app on a device. Use the app to detect the red, green, and blue of different liquids. Think about how you could use the app to detect the presence of certain chemicals in a liquid.

CH_3
O_2N
NO_2
NO_2
H_3C
O
N
CH_3

Laura Bassi

AIR/ELECTRICITY

A DEBATE

Europe's first female university professor, Laura Bassi, was born to a middle-class family in Bologna, Italy, in 1711. She was the only surviving child in her family, and although she never attended school, she was tutored in math, languages, literature, and natural philosophy from the time she was five years old. Although girls were not treated equally at that time, people recognized that she had an extraordinary mind and Laura was invited to publicly debate four professors from the University of Bologna when she was only nineteen years old.

MINERVA

At twenty, Laura became the first woman to receive a doctoral degree in science from the University of Bologna, and was made a professor at the same time. She was a celebrity. Several paintings commemorated the occasion, along with a bronze medal featuring Laura's face. People called her Minerva of Bologna, after the Roman goddess of wisdom and learning, and scholars and other travelers came from far and wide to meet her.

A LABORATORY

Although she was a professor and a member of the Italian Institute of Science, Laura was not allowed to lecture in public without the permission of men at the university. Because society would allow her more freedom to converse with other scholars as a married woman, she agreed to marry the physician Giuseppe Veratti. He had promised via letter that he would not "hinder her in her studies." She set up a laboratory in their house, where she taught and did experiments. Laura Bassi had eight children, but there were no vaccines against childhood diseases and three of them died very young.

AIR AND ELECTRICITY

As an experimental physicist, Laura Bassi was fascinated by the discoveries of Sir Isaac Newton. She was especially intrigued by his work with light and sometimes did public demonstrations of his experiments. Because she was interested in the relationship between light, air, and electricity, some of her own experiments involved studying bubbles that formed in liquids, and whether they were attracted to the walls of containers due to electrical forces.

A FORCE

Laura and Giuseppe had an "electrical machine" in their laboratory that made it possible for them to do groundbreaking research at home. They conducted dangerous experiments on atmospheric electricity at their country house, after experiments with lightning were banned in the city of Bologna. Laura authored at least twenty-eight papers in her career. Most involved Newtonian mechanics, but seven of her papers were focused on electricity. She exchanged letters with Alessandro Volta, who created the first battery. Volta was a great admirer of her work.

A CHAIR

When Laura Bassi was sixty-five, she was finally named the chair of physics at the University of Bologna. She died two years later at the age of sixty-seven.

IN TODAY'S WORLD

Static electricity is used in some factories to remove pollutants from smokestacks and pipes. Tiny particles can be charged and then exposed to a metal collection plate containing the opposite charge, which pulls them out of the air before it enters the atmosphere.

TRY THIS AT HOME!

Pour carbonated liquids, such as soda, into clear containers with different shapes. Observe how bubbles form on tiny scratches in the glass and plastic.

Lise Meitner

NUCLEAR FISSION

VIENNA

Lise Meitner was born in 1878 in Vienna, a large city that was then in the multinational state of Austria-Hungary. The third of eight children, she had four sisters and three brothers. Lise's family was Jewish, and her father was a lawyer. Although they valued education, at that time girls were not allowed to pursue an education beyond primary (grade) school. When Lise was eight years old, she kept a notebook of her observations. She wrote about the rainbow of colors she saw in a thin film of oil and about the reflection of light. It was the beginning of her love of science and math.

A TOUGH EXAM

In 1901, women were finally allowed to take an entrance exam at the University of Vienna. Cramming the eight years of education she had missed into two years, Lise was one of few women admitted to the school. In college, she developed an interest in physics and became the second woman to earn a doctoral degree at the university.

ATOMIC WOMAN

A talented researcher, Dr. Meitner studied radioactivity. Some of her experiments involved sending beams of radioactive particles into metal foil, such as aluminum foil. She was surprised to find that the higher the atomic mass of the metal in the foil was, the more the radioactive particles scattered when they hit it. Her work later helped a man named Ernest Rutherford determine the structure of atoms.

A COLLEAGUE

Lise Meitner moved to Berlin, Germany, which was called Prussia back then. She attended lectures by the famous physicist Max Planck and did research with a scientist named Otto Hahn. Women were not allowed at the research institute where she worked, so she had to work in a woodshop, which served as her laboratory, and use the bathroom at a restaurant down the street. Otto Hahn was made a professor, but Meitner was not paid for her work. After a few years, things got better. She became friends with several physicists at the institute, but she was never paid as well as Otto.

AN ESCAPE

As an adult, Lise converted to the Christian religion, but World War II put her in grave danger because of her Jewish heritage. Otto Hahn helped her escape to Sweden, where she was safe. Sadly, all her equipment was in Germany, and she could not continue her research in person. Hahn continued to send her his experimental results, which suggested that the nucleus of an element called uranium could be split.

A REVELATION

Although Hahn could not decipher how a nucleus could split apart, Lise figured it out. While walking through wintery Swedish woods with her nephew, she imagined a droplet of water splitting in half—like a dividing cell. Lise pulled out a piece of paper and did the mathematical calculations. They added up. She wrote to Hahn and they both published papers on the phenomenon, which Meitner called fission. Otto Hahn was awarded a Nobel Prize for their discovery, but Lise Meitner was left out and got nothing. Eventually she won many other prizes for her work. Dr. Meitner hoped that her discovery would be used for good, rather than for destruction.

IN TODAY'S WORLD

Nuclear reactors use uranium as fuel to create controlled fission reactions that produce energy without burning fossil fuels.

TRY THIS AT HOME!

Roll Play-Doh into a ball to represent the nucleus of an atom undergoing nuclear fission, which means splitting in two. Sculpt the ball into an elongated egg shape. Pinch the center of the cylinder and roll each half into a new ball. As you separate the halves, imagine an enormous amount of energy being released.

Katharine Burr Blodgett

THIN FILMS

LEARNING FRENCH

Katharine Burr Blodgett was born in 1898. Katharine's father, a patent attorney in Schenectady, New York, was shot to death during a robbery shortly before her birth. Her mother moved with Katherine and her brother to Paris, France, when they were young, so they could learn a second language.

SCHOOL IN AMERICA

After going back and forth between New York and Europe for several years, Katharine finally started school when she was eight years old. She did well at her studies and got a scholarship to Bryn Mawr College, where she was inspired by her math professor Charlotte Angas Scott and her physics professor James Barnes. After college, she went to the University of Chicago to get her master's degree. While she was there, Katharine studied gas masks to learn how different materials absorbed (attached to) gases.

GENERAL ELECTRIC

Katharine went to work at General Electric, where she worked in the lab of Irving Langmuir for six years before going to Cambridge University in England to get her doctoral degree. One of very few women at Cambridge, she studied with the famous physicist Ernest Rutherford and became the first woman to earn a doctoral degree in physics from the university.

THIN FILMS

Returning to General Electric, Dr. Katharine Blodgett resumed the work she had been doing with Irving Langmuir. Interested in designing techniques for applying thin films of fluids on surfaces such as glass and metal, she refined a piece of equipment called the Langmuir-Blodgett trough. The pan-like apparatus* allowed her to float oily substances on water, and she used them to coat glass and metal. Some of the layers she created were a single molecule thick.

INVISIBLE GLASS

By adding layer after layer of non-reflective coating to glass until she had forty-four thin layers, Katharine created glass with no reflection, which proved extremely useful. Her non-reflective glass was used for movie cameras, telescopes, non-reflective eyeglasses, and submarine periscopes and spy cameras during World War II.

A CREATIVE MIND

Katharine Burr Blodgett won many awards and was inducted into the National Inventors Hall of Fame. Besides non-reflective glass, she invented a "color ruler" that allowed scientists to measure the thickness of thin layers and a device to measure humidity. Katharine was a conservationist and an actress in her local theater, and she loved gardening. She also liked playing cards, writing funny poetry, and looking through telescopes. Dr. Burr Blodgett died in 1979.

IN TODAY'S WORLD

Katharine Burr Blodget's thin-film technique, which creates Langmuir-Blodgett (LB) films, is used to make computer components and anti-reflective glass, among other things.

*The design of the Langmuir-Blodgett (LB) trough was inspired by an apparatus designed by Agnes Pockels (pages 12–13).

TRY THIS AT HOME!

Fill a disposable flat container with water. Cut black paper into pieces smaller than the container. Add a drop of clear nail polish to water. Wearing gloves, quickly lay the paper on the water. Try to lift the clear, thin film of nail polish floating on the water. It may take a few tries.

1
H
2
He

Cecilia Payne-Gaposchkin

STAR MATERIAL

A LIBRARY

Cecilia Payne was born in Buckinghamshire, England in 1900. Her father died when she was only four, leaving her mother to raise Cecilia and her two siblings. The family was financially well-off and had a large library, where Cecilia read literature and learned about music and theater. They moved to London when she was twelve, so her brother could go to a better school.

AN EDUCATION

Cecilia's talent for music led the well-known composer Gustav Holst to encourage her to make a career of it. She loved science more than music, though, and she won a scholarship to Newnham College, a women's college at Cambridge University. At that time, few women studied science and they had to sit at the front of lecture halls, enduring the teasing of male students. Cecilia kept her focus and completed her studies there, although at the time, Cambridge did not award degrees to women.

STARS IN HER EYES

Cecilia became fascinated with the stars while attending a lecture by a man named Arthur Eddington. He spoke about photographing stars off the coast of Africa in 1919 during a total eclipse of the sun to test Albert Einstein's theory of relativity. She was so excited about his lecture that she almost had a nervous breakdown.

STELLAR RESEARCH

In 1923, Cecilia moved to the United States to work with the astronomer Harlow Shapley at the Harvard Observatory. She spent the next two years working day and night to analyze thousands of images from the observatory's collection of stellar spectra, which contained the light signatures from stars. Cecilia combined the data she interpreted from the collection with a new theory developed by Meghnad Saha, a scientist from India, and came up with a revolutionary theory of her own—that the sun is made mostly of hydrogen and that helium is the second most plentiful element.

A BRILLIANT THESIS

When Cecilia showed her work to Harlow Shapley and his colleague Henry Norris Russell, they rejected her conclusion, siding with the current theory that the sun had the same elemental composition as Earth. She published her thesis, but downplayed her discovery, and became the first woman to earn a doctoral degree in astronomy from Harvard. It was not until Russell came to the same conclusion himself years later that people accepted that the sun was made of hydrogen and helium. Russell was congratulated for "his" discovery, while Cecilia got no credit until decades later.

A TRAILBLAZER

Dr. Cecilia Payne married a Russian astronomer named Sergei Gaposchkin and they had three children who spent many hours in the observatory because Harvard didn't pay them enough to afford childcare. In 1956, Cecilia became the first female professor of the Harvard Faculty in Arts and Sciences and the first woman to head a department there. Besides writing five books, she has an asteroid named for her.

IN TODAY'S WORLD

Every object reflects, absorbs, or produces light waves. Spectroscopy is a technique that uses a prism or grooves to split light waves from an object, such as a star, into its different wavelengths. It is one of the most powerful tools used by astronomers today to study what stars are made of.

TRY THIS AT HOME!

Use the grooved side of a compact disc to focus sunlight onto a piece of white paper and create a rainbow. This collection of colors is called a spectrum. Spectrums are like fingerprints and the wavelength of the colors tells scientists what elements celestial objects are made of.

Mary Golda Ross

AEROSPACE ENGINEERING

GREAT GRANDDAUGHTER OF A CHIEF

Mary Golda Ross was born in Park Hill, Oklahoma, in 1908. The second of five children, she was the great granddaughter of John Ross, who had been the chief of the Cherokee nation from 1828 to 1866. Despite the best efforts of Indigenous leaders such as John Ross, the United States government had forcibly removed a hundred thousand Native Americans from their land and resettled them in Oklahoma in a deadly, devastating march from their homes known as "The Trail of Tears."

TAHLEQUAH

Mary was extremely bright, and she was happy to be brought up in the Cherokee tradition of equal education for boys and girls. Her parents sent her to live with her grandparents in Tahlequah, the capital of the Cherokee Nation, which stood beside the foothills of the Ozark Mountains. When she turned sixteen, Mary went to Northeastern State Teachers College in Tahlequah, where she earned her bachelor's degree in mathematics.

THE GREAT DEPRESSION

During the Great Depression, Mary taught math and science at a school out in the countryside. After teaching for nine years, she went to work at the Bureau of Indian Affairs in Washington, D.C. and then at an American Indian boarding school in Santa Fe, New Mexico. In 1938, Mary moved to Colorado to get a master's degree in mathematics. While she was in school there, she took as many astronomy classes as she could.

THE WAR

When World War II broke out, Mary moved to California to work for Lockheed, an American aerospace company. Her job included studying the effects of air pressure on fighter jets and water pressure on submarine-launched vehicles. Mary was so smart and talented that in 1952 she was made part of a top-secret think tank called "skunkworks" in which she was the only woman and the only Native American.

A ROCKET SCIENTIST

After the war, Mary studied aerospace engineering as she continued to work on designing satellites and rockets. She was part of the Space Race, which aimed to send humans out of Earth's orbit. At Lockheed, Mary helped write NASA's Planetary Flight Handbook, a guide to space travel. She believed that women would make great astronauts, although she added "I'd rather stay down here and analyze the data."

AN INSPIRING FIGURE

Mary Golda Ross eventually became a senior advanced systems staff engineer at Lockheed. When she retired, Mary spoke at schools, encouraging young women and Indigenous youths to consider careers in technology and engineering. At the age of ninety-six, she participated in the opening ceremonies for the National Museum of the American Indian in Washington D.C. Her image can be found on the 2019 one-dollar coin, along with that of American Indian astronaut John Herrington, whom her work helped launch into space in 2002.

IN TODAY'S WORLD

The work of Mary Golda Ross and her colleagues is engrained in today's aerospace industry, from rocket designs to satellites.

TRY THIS AT HOME!

Look up instructions on how to fold a paper airplane and test how far it will fly. Then, change the wing and tail design to see whether you can engineer that type of plane to fly farther. Try the same thing with other paper airplane designs.

Chien-Shiung Wu

SYMMETRY

A SUPPORTIVE FATHER

Born in 1912, Chien-Shiung Wu loved both her parents, but was especially close to her father, who believed that girls should have an education. She and her two brothers grew up in a small fishing village in the Jiangsu province of China. Until she turned eleven and went away to boarding school, Chien attended an elementary school for girls that had been founded by her dad.

A SEA VOYAGE

After graduating at the top of her class, Chien attended National Central University in Nanjing, where she studied math and then physics. She worked in a lab for a few years following graduation and then boarded a ship for America, with plans to attend the University of Michigan for graduate school. After she arrived in San Francisco, a young scientist named Luke Chia-Lu Yuan took her on a tour of the Radiation Laboratory established by Ernest Lawrence in Berkeley.

CHANGE OF PLANS

When Chien heard that women were not allowed to use the front entrance at the University of Michigan, she stayed in Berkeley to pursue her doctoral degree there, in the Radiation Laboratory. Chien used the cyclotron particle accelerator Dr. Lawrence had developed to study a process called beta decay. To do this, she measured electromagnetic energy, called radiation, which was given off by charged particles when they crashed into other particles. In 1940, Chien received her doctoral degree.

THE MANHATTAN PROJECT

Eventually, Dr. Chien-Shiung Wu married Yuan. Chien and Yuan moved to the East Coast, and Dr. Wu taught at Smith College and Princeton, where she was the first woman faculty member of the physics department. During World War II, she joined the Manhattan Project, which was developing the atomic bomb. In addition to improving radiation detection equipment such as Geiger counters, she helped solve a problem that was shutting down an important nuclear reactor. After the war, Dr. Wu became a professor at Columbia University in New York.

THE WU EXPERIMENT

As Dr. Wu continued to study beta decay, she was approached by two of her colleagues who were trying to prove a theory they had about how certain subatomic particles behave. They believed certain symmetrical particles that looked identical, but were mirror images of one another, did not always behave in the same way. The men, Dr. Lee and Dr. Yang, had drawn a rough sketch of an experiment out on paper, but didn't have the expertise to perform it themselves.

Dr. Wu developed a technique that allowed her to perform the experiment using strong magnetic fields, extremely low temperatures, and radioactive cobalt. She demonstrated that under the experimental conditions she had created, Lee and Yang's theory was correct. The famous experiment was named the Wu experiment, but her contribution to the groundbreaking work was ignored when Lee and Yang were awarded the 1957 Nobel Prize.

THE WOLF PRIZE

In addition to the Wu experiment, Dr. Wu confirmed Enrico Fermi's theory of beta decay and studied molecular changes in sickle cell disease. She was awarded the prestigious Wolf Prize in 1978. Dr. Wu is buried in her hometown in China in the courtyard of the school her father established.

IN TODAY'S WORLD

Dr. Chien-Shiung Wu was an activist against gender discrimination and is a role model for women in science.

TRY THIS AT HOME!

Fold a piece of paper in half. Unfold it and paint a name or design on one side of the fold. Refold the paper along the same line, smoothing it to transfer the paint to the other half. Unfold it. You have created a symmetrical image.

Ruby Payne-Scott

SUNSPOTS

A WELL-ROUNDED STUDENT

Ruby Payne-Scott was born in 1912 in New South Wales, a state on the east coast of Australia. As a child, Ruby moved to Sydney to live with her aunt. She attended grade school and high school there, winning honors in both math and biology. At the University of Sydney, she studied physics, chemistry, mathematics, and botany, graduating with a master's degree in physics in 1936.

MAGNETS AND CHICKENS

Ruby worked at a laboratory at the university for a while, studying the effect of strong magnets on developing chicks. She left to teach high school for a few years before eventually landing a job as a librarian at an electronics company that made and operated radios. They discovered her talent for research, and soon Ruby began taking measurements and doing electrical engineering.

TOP SECRET

During World War II, Ruby joined Australia's Radiophysics Laboratory, where she did top-secret work on radar technology. Before long, Ruby Payne-Scott was Australia's top expert on the radar used to detect enemy aircraft. After the war, she continued to improve radar systems and led the laboratory in a new direction—using radar to examine scientific questions.

NOISY SUNSPOTS

In 1946, the science journal *Nature* published a letter from Ruby showing a correlation between sunspots (bright flares on the sun's surface) and increased electromagnetic radio wave emissions from the sun. She and her colleagues took radio wave measurements from seaside cliff tops that confirmed her theory that sunspots produce radio bursts. Ruby designed a piece of equipment called a swept-lobe interferometer, which scanned the sky, picking up radio signals. This allowed radio astronomers to focus in on certain radio signatures that interested them.

A SECRET WEDDING

Ruby married a man named Bill Hall in 1944, but the wedding was kept secret because married women were not allowed to hold public-service jobs such as hers. When the marriage was discovered, Ruby fought to keep her job. She was fired from her permanent position, although she was permitted to keep working without a job title. When she had children, Ruby was forced to quit work entirely, because there was no maternity leave.

A CAT LOVER

When her son and daughter were older, Ruby Payne-Scott returned to radio astronomy and teaching. Besides science and her family, Ruby loved cats. She also enjoyed walking in the Australian outback and knitting. She died in 1981.

IN TODAY'S WORLD

Ruby's work in radioastronomy helped create the technology used to discover black holes and pulsars. It also helped scientists understand the effect of solar storms on electrical systems on Earth.

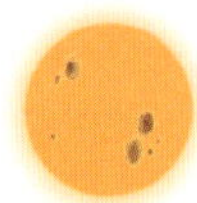

TRY THIS AT HOME!

Build a solar viewer by cutting a large notch in one end of a shoebox. Cover the notch with foil and poke a single pinhole in the center of the foil. Tape white paper to the opposite end of the box. Standing with the sun behind you, focus the sunlight coming through the pinhole on the opposite side of the shoebox to see an inverted image of the sun.

Christine Darden

AIRCRAFT WING DESIGN

A TEACHER'S KID

Christine Mann was born in 1920 in Monroe, North Carolina. Her father sold insurance. Christine's mother, a teacher, began bringing Christine to school when she was three years old. By the time she was four, Christine had started kindergarten. Soon, she discovered a passion for taking things apart and putting them back together. Breaking down and rebuilding her bike was one of her first projects.

CIVIL RIGHTS

After graduating as the valedictorian of her high school class, Christine attended Hampton University in Virginia. While she was there, she became involved in the early civil rights movement and attended some sit-ins with her classmates. She graduated in 1962 with a degree in mathematics and taught high school for a few years.

AEROSOL PHYSICS

When she was a teacher, Christine married Walter Darden and changed her last name. She got a job in an aerosol physics lab, studying the aerodynamics of how tiny particles move through the air. While working in the lab, she also taught and earned a master's degree in mathematics from Virginia State University.

A HUMAN COMPUTER

In 1967, Christine was hired by NASA to work as one of their "human computers," a group of women who solved mathematical equations for engineers working in the Space Race to safely send humans into space and bring them back to Earth. The Black women working as computers were segregated from the white women with the same jobs. To make doing the calculations more efficient, Christine wrote programs for the early computers at Langley Research Center, where she worked. She and the women she worked with played an integral role in making NASA's first space missions possible.

A GOOD QUESTION

When Christine asked the director at NASA why men were made engineers, while equally qualified women were not given the same opportunity, he was surprised. Nobody had ever asked him the question before. Christine Darden was transferred to the engineering team, earned a PhD in engineering from George Washington University, and was the first African American appointed to the highest rank at Langley Research Center.

SONIC BOOM

In her role as the leader of Langley's Sonic Boom Team, she used computer programs to test aircraft designs. One of Dr. Darden's main jobs was to redesign super-fast aircraft to reduce the shock waves they can cause in the atmosphere. The name *sonic boom* refers to the thunderous sound made by air molecules crashing together when an object moves faster than the speed of sound. She also tested models of airplanes in wind tunnels to study their aerodynamics.

HIDDEN FIGURE

Christine Darden retired in 2007, but she continues to serve as a role model for young mathematicians, scientists, and engineers. She advises aspiring engineers to picture themselves in their dream job, plan, prepare, and persist. Her story was featured in the film *Hidden Figures*.

IN TODAY'S WORLD

NASA is still working to design supersonic aircraft that produce less noise and environmental impact.

TRY THIS AT HOME!

Look up instructions for how to fold a paper airplane and test how far it will fly. Then, change the wing and tail design to see whether you can engineer a plane that will fly farther. Try the same thing with other paper airplane designs.

Valerie L. Thomas

ILLUSION TRANSMITTER

A SELF-STARTER
Valerie L. Thomas was born in Maryland in 1943. She first became interested in science while watching her father fix their television set. Valerie checked a book called *The Boys' First Book of Electronics* out of the library. Although her father refused to help her with the projects in the book, Valerie learned what she could on her own.

PHYSICS CLASS
Valerie's family and most of her teachers were not supportive of her interest in science and engineering, but she took a physics class in school. That class, along with the encouragement of a few teachers, kept her interested in the sciences, and she went to college to study physics at Morgan State University, where she was one of only two women majoring in physics.

NASA
After graduating with a physics degree, Valerie went to work at NASA, analyzing data from satellites. She developed computer data systems and oversaw the development of LANDSAT, the longest-running program for collecting satellite data from Earth. Data from these satellites allowed Valerie and her coworkers at NASA to create a system for making predictions about world-wide food crop yields.

ILLUSION TRANSMITTER
An exhibit at a scientific seminar in 1976 caught her attention. Using mirrors, the device gave the illusion that a lightbulb was still glowing, even after it had been turned off. Fascinated, Valerie experimented on her own, using mirrors to produce three-dimensional (3-D) images in front of the glass. In 1977, she used what she'd learned to invent a device called the illusion transmitter. The device, which she patented, was the forerunner of the 3-D technology we use today.

PROBING SPACE
While at NASA, Valerie Thomas became the associate chief of the Space Science Data Operators office. She helped develop technology for NASA's scientific network and the modern internet. Valerie was also involved in projects related to the Voyager Spacecraft, a space probe studying the outer solar system and Halley's Comet.

A ROLE MODEL
Valerie was awarded the Goddard Space Flight Center Award of Merit and NASA's Equal Opportunity Medal. During her career at NASA, and following her retirement in 1995, she made hundreds of school visits, encouraging minorities and women to pursue science and engineering careers.

IN TODAY'S WORLD
Valerie Thomas's ideas are still being used to create modern 3-D technology. The LANDSAT satellite program she created monitors Earth's atmosphere and oceans, tracking climate change. Data coming from these satellites aid the government in everything from land management to helping endangered species.

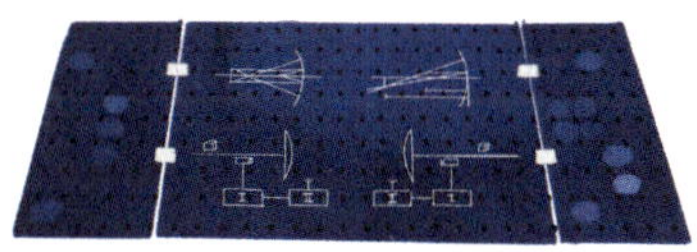

TRY THIS AT HOME!
Look at your reflection on the back of a spoon. Flip the spoon around so that the image is flipped upside-down (inverted). Your image is inverted because of the way light is reflected by curved, concaved surfaces. This physics can be used to create useful optical illusions.

• PHYSICS •

Jocelyn Bell Burnell

PULSARS

SPUTNIK

Jocelyn Bell Burnell was born in Lurgan, Northern Ireland, in 1923. Her father sparked her interest in the stars by checking out astronomy books from the library and waking Jocelyn and her sister to watch the satellite Sputnik pass overhead. When Jocelyn was thirteen, she went to boarding school in England, where her physics teacher Mr. Tillott encouraged her interest in science.

A BOY'S CLUB

As the only woman studying physics at the University of Glasgow, Jocelyn was harassed by the male students. They whistled, stomped their feet, and banged their desks each time she entered the lecture hall. Jocelyn graduated with honors and a degree in physics, but the experience of feeling so alone stayed with her. From there, she went on to Cambridge University.

A WINDOW TO THE UNIVERSE

Radio physics was a new and exciting field. Bursts of electromagnetic energy from the far reaches of the universe could be detected and recorded as radio signals that made bumps and squiggles on the paper of recording instruments. The farther away objects were, the older they were, so studying far-off objects was like traveling back in time.

SIGNALS FROM SPACE

Jocelyn helped her graduate advisor Anthony Hewish construct a radio telescope that covered four acres and looked like an agricultural frame made of bean poles. While strong signals from deep space registered as strong peaks in the data flowing from the telescope, there were also background radio signals from human activity. Jocelyn was put in charge of running the telescope.

A SQUIGGLE

Every night, Jocelyn went to the university's observatory to analyze signals coming from the radio telescope. She studied 900 feet of paper each day, searching for peaks representing electromagnetic bursts coming from space. One night in 1967, she noticed a funny squiggle on the paper. Similar peaks appeared on the paper several times at regular intervals, indicating rhythmic pulses of energy from deep space.

Jocelyn contacted Dr. Hewish about her discovery, but he did not believe her until he observed the signal himself and they found it on a second telescope. Soon afterwards, she stumbled uninvited into a meeting in which Dr. Hewish and other men were discussing (without her) the origin of the signal she had discovered.

PULSARS

Later that night, Jocelyn marched back out to the observatory in the freezing cold to point the radio telescope at a different part of the night sky, where she recalled seeing a similar signal. She found it immediately, and the second pulsing signal confirmed that the radio waves were not background noise, but came from an entirely new type of star, which they later called a pulsar.

Dr. Jocelyn Bell Burnell's discovery of pulsars changed our understanding of the universe. She was left out, however, when Hewish and Martin Ryle were awarded the 1974 Nobel Prize for Physics. In 2018, she was awarded the Breakthrough Prize and three million dollars, which she donated to the Institute of Physics to increase diversity in the field.

IN TODAY'S WORLD

Pulsars are important tools for physicists, and they are used to test theories of relativity on a cosmic scale.

TRY THIS AT HOME!

Make a model of a pulsar by taping two flashlights pointing in opposite directions onto a ball. Turn the flashlights on and go into a dark room. Spin the ball so the flashlights make a constant, repeating pattern on the wall, representing the radio signals Jocelyn found on her telescope.

Nadya Mason

CONDUCTIVITY/CARBON

POKING AT ANT HILLS

From the time she was young, Nadya Mason has been interested in the world around her. As a child, she loved nature and satisfied her curiosity by running around picking fruit from trees and poking at anthills with sticks. Born in New York City, she lived in Brooklyn until she was six. Her family then moved to Washington D.C. and later to Houston, Texas.

A WORLD-CLASS GYMNAST

An exceptionally talented athlete, Nadya ran track and was a gymnast on the United States National team. In addition to sports, she was an excellent student who became interested in math at a young age. It wasn't until she had an opportunity to work in a biochemistry lab that she discovered her love of hands-on science and realized for the first time that it was possible to make science her career.

• PHYSICS •

ATOMIC AMBITIONS

After high school, she went to Harvard University, where she took as many science and math classes as she could. Soon Nadya discovered that she likes physics the best. The science of atoms, the building blocks of matter, described the world in a way she understood and connected with. Her parents were supportive of the idea that Nadya would be a physicist and encouraged her to pursue her dreams.

A PROFESSOR

After getting her bachelor's degree at Harvard, she continued her studies at Stanford University, where she received her doctoral degree. Dr. Mason returned to Harvard to do some post-doctoral research before becoming an assistant professor and then a full professor at the University of Illinois Urbana-Champaign. Besides doing research there, she loves working with her students.

ELECTRON INTERACTIONS

Dr. Mason's laboratory focuses on how charged particles called electrons move through different materials, how they interact with each other, and why they conduct electricity the way that they do. She's especially interested in how electrons behave in very thin layers or small quantities of different materials. Nadya explains, "we work with carbon nanotubes that are just one-billionth of a meter in diameter but up to centimeters long; think of them as the tiniest wires you can make."

HANDS-ON THINKING

Dr. Mason is passionate about diversity issues that affect the scientific community. She works hard as a science communicator to encourage everyone to follow their interests in science, rather than being intimated by it. Dr. Mason gave a TED talk in 2019 titled "How to spark your creativity, scientifically," and said that "hands-on thinking (through experimentation) connects our understanding and even our vitality to the physical world and the things that we use." She has won several awards, including the American Physical Society's Maria-Goeppert Mayer Award, and was a general counselor to the American Physical Society (APS) and chair of the APS Committee on Minorities.

IN TODAY'S WORLD

Dr. Mason's research on electron movement through very small quantities of material could have an enormous impact on modern technology, such as the computer industry.

TRY THIS AT HOME!

Use a graphite pencil to scribble a solid rectangle on some paper. Stick a piece of clear wrapping tape to the graphite and lift the tape up. With a second piece of tape, lift graphite from the first piece of tape. Repeat several times with new pieces of tape, adhering them to white paper after lifting graphite. Photograph the tape. Zoom in to observe the successively thinner layers you made.

Chanda Prescod-Weinstein

COSMOLOGY

A BASEBALL FAN

Chanda Prescod-Weinstein was born in the El Sereno neighborhood of Los Angeles, California, where the Tongva village of Otsungna stood until Spanish colonizers arrived around 1770. El Sereno is near Dodger Stadium, and as a kid Chanda became a huge fan of the Dodgers baseball team. In addition to dreaming of becoming a professional baseball player, she played piano, flute, and saxophone.

READING, DANCING, AND MATH

When she was young, Chanda loved practicing her multiplication tables. Soon, she became comfortable with mathematics, which is the language of physics. In high school, books such as Stephen Hawking's *A Brief History of Time* got her excited about black holes and the building blocks of the universe. In addition to reading and studying, Chanda attended a performing arts school, where she trained as a dancer in both modern dance and jazz.

FRESHMAN PHYSICS

Chanda attended Harvard University, where she earned degrees in physics and astronomy. Despite finding her freshman physics class somewhat boring, she later realized that the simple experiments they did in class built a solid foundation that would help her solve more interesting and complicated problems in the future.

A MENTOR

Most of Chanda's physics professors and classmates were white men, so she had to create her own vision of herself as a physicist. When she was almost done with her undergraduate degree, she finally met a Black woman with a doctoral degree in physics— Dr. Nadya Mason, who became a role model and a mentor. Chanda got a master's degree in astronomy, a doctoral degree in theoretical physics, and a job as an assistant professor in physics and core faculty in women's and gender studies at the University of New Hampshire.

UNIVERSAL QUESTIONS

As a theoretical physicist, Dr. Prescod-Weinstein uses math to solve the mysteries of the universe. Scientists think that the universe contains a lot of invisible stuff, called dark energy and dark matter, along with "a tiny smidgeon of everyday stuff like us." Dr. Prescod-Weinstein uses ideas from physics and astronomy to learn more about dark matter and address questions about how everything in the universe "got to the be the way it is."

AN AUTHOR AND AN ACTIVIST

Besides studying particles physics, Dr. Prescod-Weinstein is a popular science communicator and an activist for equality in science. She is a Pilates instructor, loves *Star Trek*, and does research on feminist science studies. Her book *The Disordered Cosmos: A Journey into Dark Matter, Spacetime, and Dreams Deferred* explores these ideas. She has received several prestigious awards for her work in particle physics, astrophysics, and astronomy.

IN TODAY'S WORLD

Scientists are currently using data collected by ground-based and space observatories to study the light coming from 35 million galaxies. They hope to reconstruct how the universe has been expanding over the last 11 billion years. Ideally, this work will help us understand the nature of dark energy.

TRY THIS AT HOME!

Make a batch of bread dough or pizza dough. Add poppy seeds, sesame seeds, or spices such as oregano to the dough to represent galaxies. As the dough rises, the seeds and spices will move apart. This shows how as the universe expands outward in all directions, galaxies and other cosmic objects move away from each other.

Burçin Mutlu-Pakdil

GALAXIES

ISTANBUL

Burçin Mutlu-Pakdil was born in the Turkish city of Istanbul, which lies in both Europe and Asia. Istanbul is an ancient, famous city that went by the names Byzantium and Constantinople in the past. Burçin's grandparents never learned to read or write and her parents were not able to pursue an education beyond the fifth grade due to financial issues. Despite those challenges, her parents believed strongly in education.

REACHING FOR THE STARS

In middle school, Burçin became interested in physics when she read a book about Albert Einstein. She had always liked looking at the stars, and now she wanted to learn more about them. Her family encouraged her to study physics, and despite the disapproval of some friends and relatives, Burçin left Istanbul to attend Bilkent University in Ankara, Turkey.

FOCUSING ON PHYSICS

When Burçin arrived at college, one of her male professors told her it was crazy for a woman to leave her hometown to study physics. At the university, she was not allowed to wear her hijab and constantly battled prejudice. Despite all the obstacles, she refused to let anything stand in her way and remained focused on her passion until she graduated. Burçin then moved to the United States to study astrophysics, earning a master's degree from Texas Tech and a PhD from the University of Minnesota. Today, Dr. Burçin Mutlu-Pakdil studies the most mysterious objects in the universe, at the University of Chicago.

BURCIN'S GALAXY

Burçin and her research team first spotted a very peculiar object in the background while studying another galaxy. At first glance, it was a beautiful celestial object with a ball of yellowish stars surrounded by a single ring of blue stars. Galaxies like these are extremely rare. When looking at the galaxy in different lights, they unexpectedly discovered an additional unique structure—a second inner ring, making this galaxy the first example of an elliptical galaxy with two symmetric rings. This unique, beautiful elliptical galaxy, 359 million light years away, is now commonly called Burçin's Galaxy.

UNIVERSAL MYSTERIES

Dr. Burçin Mutlu-Pakdil uses some of the largest telescopes in the world. She travels to locations such as Hawaii and Chile to collect data and study scientific mysteries, such as dark matter and how the faintest and smallest galaxies form and evolve in the universe. In addition to having a galaxy named after her, she has won many awards for her work in physics and is a role model for young scientists. She says, "I do not want to blend in; I want to stand out as stars do, so I fought against all these stereotypes and worked hard to live beyond the labels."

IN TODAY'S WORLD

Our galaxy, the Milky Way, formed around 14 billion years ago, attracting heavy elements that came together to form the sun, and our planet. Using powerful tools such as the Keck telescopes in Hawaii, astronomers can see the light from 100 billion other galaxies. Each of these galaxies contains 100 billion stars. By using spectra collected by the telescopes, astronomers are discovering what the galaxies are made of. They hope to learn more about the universe and whether there may be other galaxies containing planets that could support life.

TRY THIS AT HOME!

Tie a Wiffle ball onto either end of a 3-foot (0.9-m) cord or thin rope to represent two galaxies connected by gravity. Cover the balls with duct tape to make them heavier. Holding the galaxy model by one end, swing the rope a few times and let it fly through the air to see how two galaxies might move together through space.

ABOUT THE AUTHOR

Liz Heinecke has loved science since she was old enough to inspect her first butterfly. After working in molecular biology research for ten years and getting her master's degree, Liz left the lab to kick off a new chapter in her life. Soon she found herself sharing her love of science with her three kids, journaling their science adventures on KitchenPantryScientist.com. Her desire to spread her enthusiasm for science to others soon led to regular TV appearances, speaking engagements, and her books: *Kitchen Science Lab for Kids* (Quarry Books), *Outdoor Science Lab for Kids* (Quarry Books), *STEAM Lab for Kids* (Quarry Books), *Star Wars Maker Lab* (DK), *Kitchen Science Lab for Kids: Edible Edition* (Quarry Books), *The Kitchen Pantry Scientist: Chemistry for Kids* (Quarry Books), *The Kitchen Pantry Scientist: Biology for Kids* (Quarry Books), *Sheet Pan Science* (Quarry Books), *The Padawan Cookbook* (Insight Editions), *The Kitchen Pantry Scientist: Ecology* (Quarry Books), and *RADIANT: The Dancer, The Scientist, and a Friendship Forged in Light*, an adult nonfiction narrative about Marie Curie and Loie Fuller (Grand Central Publishing). Most days, you'll find Liz at home in Minnesota. She graduated from Luther College with a BA in art and received her master's degree in bacteriology from the University of Wisconsin, Madison.

ABOUT THE ILLUSTRATOR

Kelly Anne Dalton is an artist, illustrator, and storyteller living in the wild mountains of Montana. Her elegant and enchanting work can be found on everything from board books to middle grade novel covers, home decor and gift products, and stationery lines. When not drawing, daydreaming, and creating new stories and characters, Kelly Anne can be found trail running in the forests near her home.

ACKNOWLEDGMENTS

I'm extremely grateful to each of the helpful, talented people who collaborated to put this book together: my long-time acquiring editor Jonathan Simcosky; art designer Heather Godin, editorial project manager Gabrielle Bethancourt-Hughes; editorial intern Gabrielle Cruz, design intern Meigan English, marketing managers Angela Corpus and Mel Schuit; and the entire design and editing team. I'm lucky to have such a skilled, supportive group to work with. Thank you to Kelly Anne Dalton, who brings the scientists in these pages to life so beautifully, and to my literary agent Victoria Wells Arms. Special thanks to each of the scientists who took time from their busy schedules to answer my questions and tell me stories from the field. And a shout-out to my family and friends, who keep me laughing and learning every day.

ALSO AVAILABLE

ISBN: 978-1-57715-674-1

ISBN: 978-1-57715-675-8

ISBN: 978-1-57715-672-7